# Welcome to Tomorrow

An Introductory Handbook to Artificial Intelligence
and its Impact on Society

by Shanka Jayasinha

# Contents

# Introduction

**"Artificial Intelligence will replace humans!"**

How many times have you heard this in the news? Or being portrayed in Hollywood movies by robots taking over the world?

The truth is AI (Artificial intelligence) will not take over the world... At least not anytime soon.

Firstly, no one can clearly define AI, even researchers argue about an exact definition. Secondly, we are extremely far from this situation from a technical perspective. AI will mostly serve as a complementary tool, however it will definitely take over certain tasks and push people out of certain categories of jobs. Hopefully, this should be beneficial for society as businesses integrating AI will be more productive than the others and this will enable people to accomplish more and/or focus on tasks that cannot be done by computers, which should boost innovation and therefore, the economy forward.

This handbook is an attempt to explain the concept of Artificial Intelligence, its potential impact and limits in the simplest way possible without losing the readers who hate mathematics & coding!

Let's start with defining the concept of AI.

# Chapter 1: What is AI?

AI scientists do not have a clear definition of what AI is. Some people believe AI are human-made computers that can match and/or surpass actions taken by humans while others believe that any system processing any kind of data is AI. The truth is that AI keeps evolving and so does its meaning. Certain tasks that are considered AI today will be pushed back to simple fields such as statistics, probability or else once they are well understood in a near future. So what could be a timeless definition of AI?

To give you some perspective, let's look at human intelligence. It's the ability humans have to gather information, put it through cognitive processes and react accordingly. Cognitive processes correspond to the "acquiring (of) knowledge and understanding through thought, experience, and the senses"[1]. Artificial Intelligence is the same, except that this is done by a computer, hence the use of the word "artificial". Therefore, AI systems are autonomous and do not need a human to operate them constantly.

If this is broken down in the simplest form, we find three major categories: Input (Perceived Information), Interpretation (Data Processing) & Output (Decision).

**Input** corresponds to the gathering of information. Humans have five senses to gather information. This input of information can be visual, auditory, gustatory, olfactory and somatosensory or a combination of some or all. As every human is different, these senses can be more or less effective, some people will see better than others while others will have a heightened sense of

---

[1] "cognition" - definition of cognition in English from the Oxford dictionary". www.oxforddictionaries.com. Retrieved 2019-07-02

taste. Therefore, they will interpret different types of information differently. Computers work the same way except that they gather data that needs to be in its simplest form.

Let's use vision as an example because humans are primarily visual beings on average. (If you do not agree, imagine if the whole population was suddenly impaired of one sense at a time, which missing sense would create the most chaos?) When you see a blue car, you are actually seeing how this object with a specific shape and presence in your environment reflects light. Your eye is capturing electromagnetic radiation with wavelengths in the 400 nm range. Therefore, the AI system will need a sensor, which plays the same role than the eye for the human. This sensor will capture wavelengths within the visible spectrum range to be able to see an object like a human. For auditory information, you would need a sensor that measures sound waves/ vibrations, for gustatory and olfactory information, you would need a sensor that can identify chemicals and for somatosensory information, you would need a pressure sensor for example. These are the simplest sensors I can think of, a myriad of them exist and their settings can be crucial in their efficiency to gather data.

To continue with our initial example, the eye transforms the wavelengths it captures into an electric signal that travels through your nerves up to your brain. In other words, the eye sends a message to the brain which then interprets it (data processing). For the computer, the sensor will need to translate the wavelength into readable data, which is always text or numbers.

**Interpretation** is the phase where your brain analyses the captured information and tries to make something out of it by comparing it to previous experiences and knowledge. For a computer, the data it receives is sent through an algorithm. The complexity of this part for people trying to understand and create AI systems resides in the decomposition of a simple action into numerous tasks. For example, when you show the blue car to another human and ask what it is, he/she will quickly say it's a blue car but how did they get to that conclusion?

A series of small tasks took place when the brain received the information from the eye. In an oversimplified breakdown, we could say that the brain identified the color as 'blue' because it was taught since its early years that this specific wavelength corresponded to blue. It then compared this shape to all the shapes it had previously seen and concluded this to be a motorized vehicle. The brain then categorized this object as a car because it had 4 wheels instead of 2. A computer would use a similar type of process to identify the object as a blue car and this is where it gets complicated.

A computer needs clear data to be able to interpret information. In order to interpret information correctly, its algorithms need to ask the "right questions" in the "right order" to be efficient and find the **output** we want. This is how simple a computer is. Now, let's add AI. Above, I mention that the color is recognized because it was taught in its early years what it was and that the shape and object are finally recognized because it compared it to previously seen shapes & objects. AI uses previous experiences stored in its database to make its decisions. This is one key difference between systems that are considered to be AI and those that are not.

Based on our previous example, what do you think is the hardest to do for an AI system: identifying an object as a car or playing a chess game? If you thought playing a chess game would be more complicated, you're wrong. A computer can calculate all the outcomes of a chess game in a split second, a simple algorithm would do as this game has a clear set of rules with well-defined boundaries and mathematical outcomes. Contrary to playing chess, identifying an object has no boundaries, an object can have an infinite amount of shapes, size, color, etc... making it more difficult to correctly identify it. As of today, no AI system can correctly identify an object over and over again[2] whereas computers have beat humans in chess since 1997![3] Imagine how

---

[2] Lowensohn, J. (2015, May 14). Wolfram has created a website that will identify any image you throw at it. Retrieved July 3, 2019, from https://www.theverge.com/2015/5/13/8603531/wolfram-image-identification-site-trained-by-chewbacca

[3] I. (n.d.). Deep Blue. Retrieved July 3, 2019, from https://www.ibm.com/ibm/history/ibm100/us/en/icons/deepblue/

complicated tasks can get when you have to build an AI system for a self-driving car?

A self-driving car will need several AI systems. It will need a navigation system to map an itinerary which will then be constantly compared to traffic and other potential disturbances such as road deviations, weather changes in order to find the best route from one location to another.[4] Another system which needs to make decisions based on the identification of static and dynamic obstacles will also be needed.[5] Each of these systems (amongst many others) must work in a synchronized manner with a high degree of precision in order to avoid accidents and bring its passengers to their destination in a timely manner.

Any physical machine that has the same ability to assess its environment with sensors, process the information through algorithms and react accordingly, such as a self-driving car, is a robot. Robotics is one of the two main fields using all types of AI and I believe, is one of the main reasons behind this AI craze as it would improve productivity (this will be discussed further in Chapter 2 & 3) and give significant advantages to any company using it.

AI is not only used in physical machines such as robotics, it can also be virtual and be present on software only. We encounter these types of AI systems every day. Whether it's Netflix, Spotify or Facebook, the majority of content that is recommended to us is personalized by AI virtual systems.[6] They can also create, analyze and alter different types of data from numbers to visual content amongst others (clearer examples will be discussed in chapter 2).

I hope I haven't lost you in this brief introduction.

In simple terms, AI systems can be defined by their ability to operate on their own to a certain extent, therefore making them autonomous combined with their ability to adapt according to their acquired experience. This is where

[4] Kavraki, L., Svestka, P., Latombe, J., & Overmars, M. (1996). Probabilistic roadmaps for path planning in high-dimensional configuration spaces. *IEEE Transactions on Robotics and Automation,12*(4), 566-580. doi:10.1109/70.508439

[5] Khatib, O. (1986). Real-Time Obstacle Avoidance for Manipulators and Mobile Robots. Autonomous Robot Vehicles, 5(1), 90-98. doi:10.1007/978-1-4613-8997-2_29

[6] Safian, R., & Safian, R. (2018, September 11). 5 lessons of the AI imperative, from Netflix to Spotify. Retrieved July 3, 2019, from https://www.fastcompany.com/90234726/5-lessons-of-the-ai-imperative-from-netflix-to-spotify

matters get a little complicated. AI regroups multiple subfields such as Machine Learning (ML) and Deep Learning (DL) which all fall under umbrella of Computer Science (CS) and Data Science (DS):

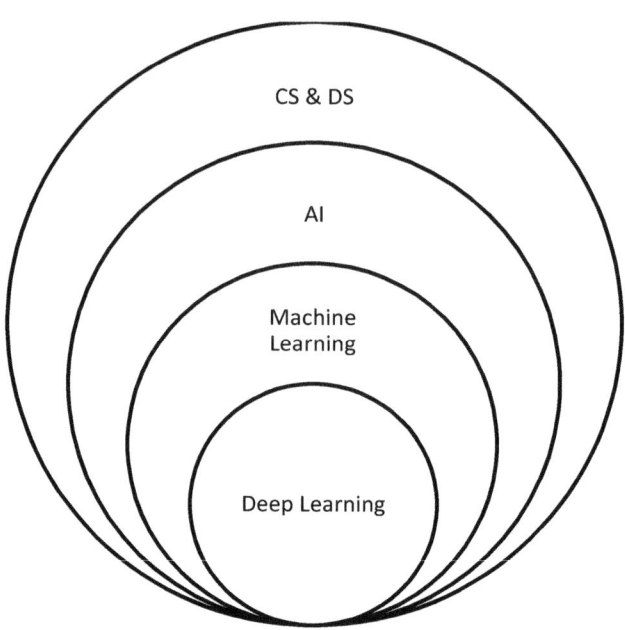

Although AI falls underneath CS & DS, these two fields are completely different from one another (another book could be written about this). In a nutshell, Computer science focuses on the development of computers as a tool, therefore touching on their data structure, programming language and overall architecture, whereas, Data Science focuses on the analysis of big data to identify patterns, trends, etc. Imagine statistics on steroids. Basically, CS develops the tools for DS, which can then focus on finding meaning in the data.[7] AI is used in both CS & DS. For example, in CS, a smart tool that can adapt to its user's needs by shifting its structure will be much more efficient than a

---

[7] Data Science vs. Computer Science. (n.d.). Retrieved July 4, 2019, from https://www.discoverdatascience.org/articles/data-science-vs-computer-science/

computer that cannot do this. Therefore, the integration of AI is a necessity, especially Machine Learning. ML can also be seen in DS as it will be useful for data mining, which is the analytical process of running different statistical models to find hidden patterns in large volumes of data.[8]

## So what exactly is Machine Learning?

I previously used the words autonomous & adaptive in our definition of AI. Well, ML systems are capable of making predictions/decisions "without being explicitly programmed to perform the task"[9] through pattern recognition and inference. Let's decompose this.

When I mention the fact that they are not explicitly programmed, it simply means that compared to traditional programming, where you have to set rules first and then send data through to obtain answers (figure 1). ML systems work in the opposite manner.

Figure 1: Traditional Programming Workflow[10]

They use data (training data) and answers to come up with rules (figure 2). These rules constitute a model, which is then used to analyze test data in order to make predictions.[10]

8    Olson, D. L. (2006). Data mining in business services. *Service Business,1*(3), 181-193. doi:10.1007/s11628-006-0014-7

9 Bishop, C. M. (2016). PATTERN RECOGNITION AND MACHINE LEARNING. Place of publication not identified: SPRINGER-VERLAG NEW YORK.

10 Moroney, Laurence. "Machine Learning Zero to Hero." I/O 2019. Google I/O 19, 9 May 2019, Mountain View, CA, Shoreline Amphitheatre.

Figure 2: Machine Learning Systems Workflow[10]

To make the dichotomy between the two types of data clearer, imagine a financial analyst, who would like to "predict" where the prices of certain stocks are going. He will build a model based on past data of stock prices, volatility, peers' performance and whatever criteria makes sense to him for his forecast. Before running his model to forecast future variations, he should test his model. One of the easiest ways is to split his past data into multiple samples and test the model on periods of time where you already know what happened to see if your model can predict the upcoming variations. This would be considered training data. If it doesn't work, the analyst needs to understand why and fix the problem. He then needs to repeat the process until his model can accurately predict upcoming variations on the overall data set, which is extremely tedious and time-consuming, especially if you consider the fact that the manual changes to the model based on one sample of the data may not fit the rest of the data set. Once the model is ready, the analyst can now try to predict where the markets are heading with the new data he gathers and adds to his model daily based on his target stocks or on another basket of stocks that share similar characteristics with the test data stocks.

However, today with ML systems, this whole process can be accomplished in a few minutes or hours depending on the size of the data. ML will find all the statistical possible relationships using different learning algorithms types.[11] [12]

[11] Oladipupo, T. (2010). Types of Machine Learning Algorithms. New Advances in Machine Learning. doi:10.5772/9385

[12] Ayodele, T. O. (2010, February 01). Types of Machine Learning Algorithms. Retrieved July 4, 2019, from http://www.intechopen.com/books/new-advances-in-machine-learning/types-of-machine-learning-algorithms

Three main types of learning algorithms exist: supervised learning, reinforcement learning and unsupervised learning.

**Supervised learning** consists in giving an input to the ML system on which it will then base itself to predict an output. For example, let's take identifying the face of a coin. The goal of the ML model would be to classify what is "Head" and what is "Tails". The input would be a picture of one of the sides, each picture would be a labeled "Heads" or "Tails". Based on all the pictures and labels the ML system has seen, it will try to set rules to make a prediction and give the right output: "Heads" or "Tails".

**Reinforcement learning** is like supervised learning except that in this case, the output is given before the input and then the system adjusts accordingly. Going back to our example of heads or tails, imagine the system being fed with a photograph without the label. It will try to predict an output and then the label would be given for it to make any necessary adjustments.

**Unsupervised learning** is the vastest learning algorithm type amongst the three. It has no outputs or any specific instruction. Although the coin example is not ideal in this case, pictures would be shown with no label and the system would try to identify patterns such as pixel brightness, color, etc. Since this is a binary classification problem, unsupervised learning would be considered overkill and useless as we already know that there are only two distinct possibilities: "Heads" or "Tails". This type of learning would be better for the financial analyst in our previous example as the system would try to cluster information.

Clustering is the process of grouping elements with similarities together.[13] There are many clustering techniques that would require some mathematical/statistical explanations, therefore, I have explained the overall idea in the chapter *To Go Further*. In a nutshell, this process would decompose the structure of the data. For our financial analyst, clustering would unveil trends & correlations, which tell him which stocks move up & down together or

---

[13] Ullman, S. (n.d.). *Unsupervised Learning: Clustering*. Lecture presented in Massachusetts Institute of Technology, Cambridge. Retrieved July 4, 2019, from http://www.mit.edu/~9.54/fall14/slides/Class13.pdf

which ones move inversely giving him valuable information on predicting future variations.

With complex tasks such as predicting stock prices, multiple layers of Machine Learning processes are required. This then becomes Deep Learning (DL). These layers are organized into a specific network named "Artificial Neural Networks" (ANN) or simply "Neural Networks" (NN). These neural networks work the same way as the ones in our brains. They process information through nodes that receive and send signals to each other. In our brain, these nodes correspond to neurons. Alone, they are not capable of much but put together, you can obtain an extremely complex network that can adapt to the different signals it receives.

There are different types of layers, the Input Layer, the Hidden Layer(s) and the Output Layer. The Input layer will decompose the information accordingly to the data size. There will be as many nodes as there are measurable units of what's being analyzed by the system. For example, in a pattern recognition system where pictures will be inserted, you could set one node per pixel. The hidden layers are the layers where the information will be decomposed and processed. There are usually as many layers as there are characteristics. If we keep the previous example, there could be a layer to measure brightness, another to measure color... Finally, you have the output layer, where the findings of the system will come out. In a simple ML system, a neural network will be straight forward (Figure 3).

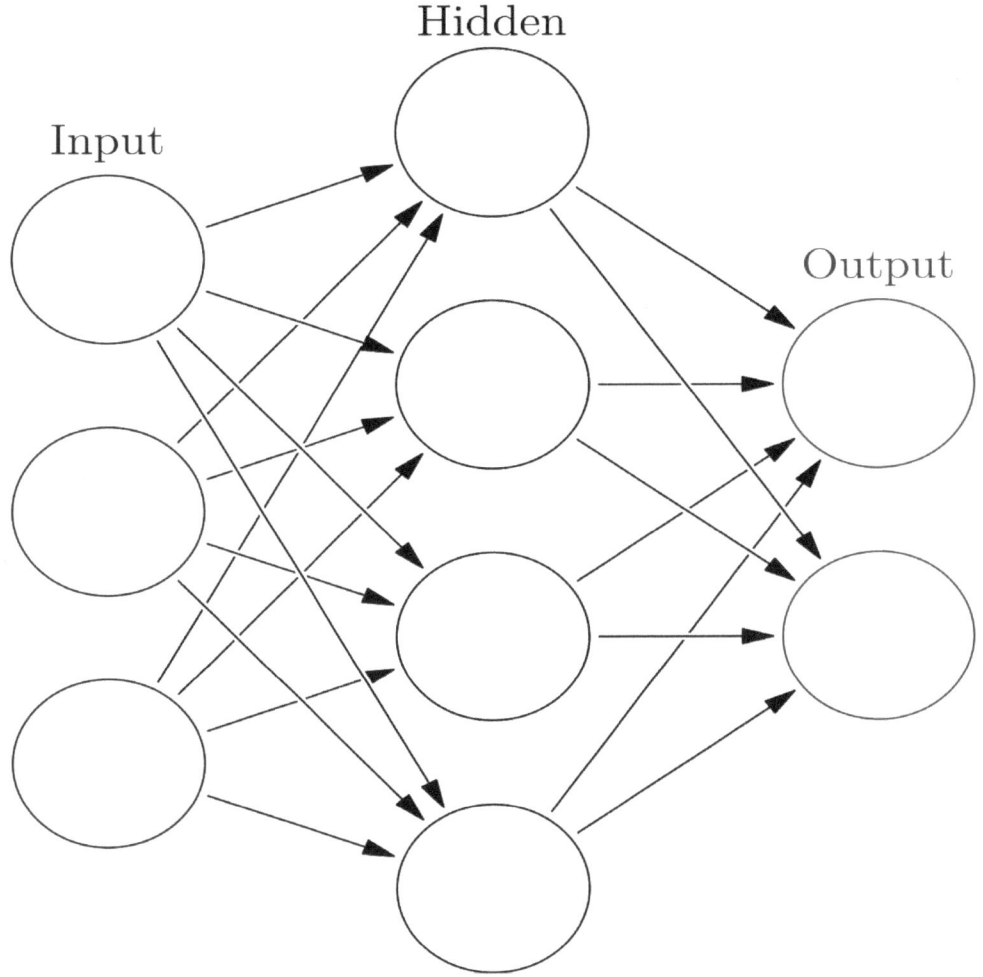

Figure 3: Illustration of a Neural Network in a simple ML system[14]

---

[14] Glosser CA. (n.d.). Colored neural network [Digital image]. Retrieved July 07, 2019, from https://commons.wikimedia.org/wiki/File:Colored_neural_network.svg

A neural network for deep learning will have multiple layers and therefore, will be a much more complex network than the previous illustration:

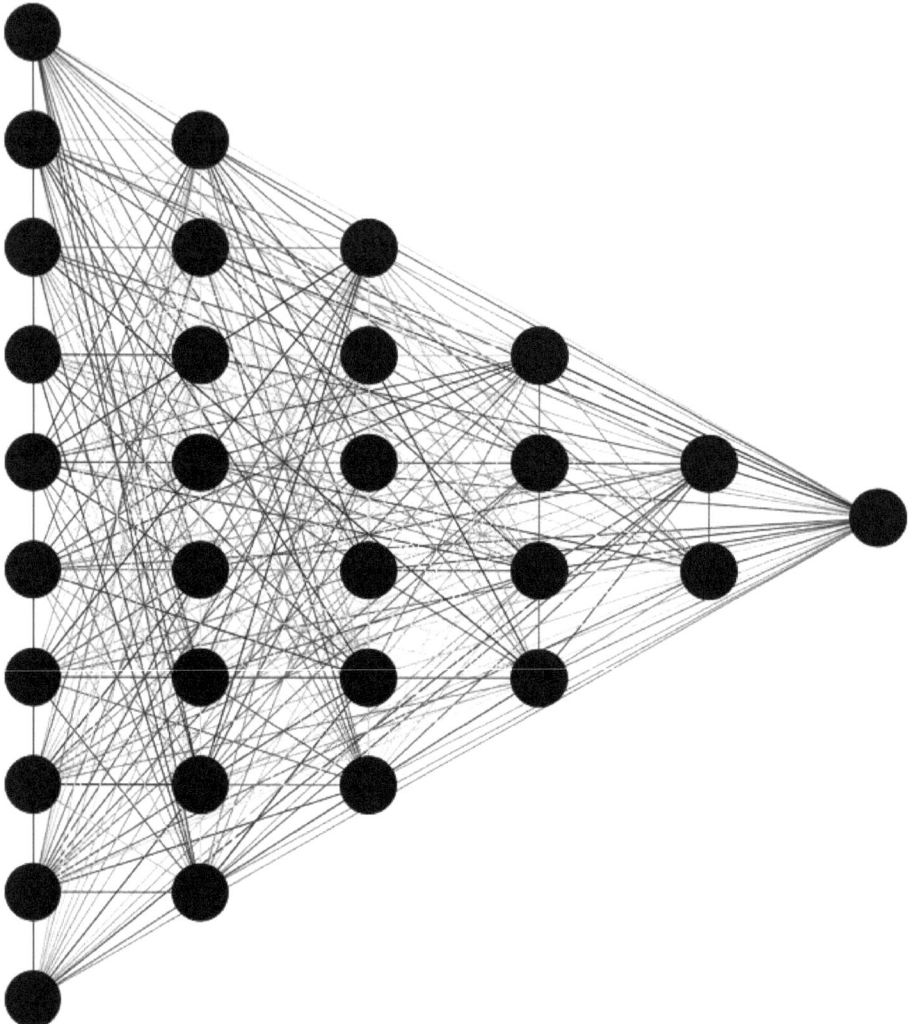

Figure 4: Illustration of a Neural Network in a Deep Learning System

In Neural Networks, every node processes the information through different levels of activity (intensity of activation), which are usually scored between 0 & 1.

Depending on the level of activity, the system can recognize different features as the information goes through different layers to give an output. Let's take a specific example. In a visual ML system that recognizes patterns, one of the layers measures brightness. If a node in that layer scores an activity level over 0.5, the concerned pixel is considered as containing an element and under 0.5, the pixel is considered as empty. Under 0.5, the network would discard the pixel for further processing as it would not be relevant. Over 0.5, it would keep processing the pixel's information through the different layers that could measure color, saturation... to see whether it is relevant or not. The whole power of neural networks resides in the fact that this process can be done with an almost infinite amount of parameters depending on the data you are giving it, the way you want the network to process it and can be applied to any sector. Therefore, although this is an artificial system, it will be exposed to a certain level of bias similarly to humans (For more details, please refer to the *To Go Further* chapter).

As our crash course on the concept and subfields of AI comes to an end, some of you may be asking why these set of processes are taking the center stage today. Let's dive into why AI has become so popular since of late.

# Chapter 2:    Why does it matter so much?

To understand why AI is so popular today, we need to take a trip down memory lane to see how it all started.

The philosophical concept of AI can be traced back to Greek myths[15], where it was believed that placing human thinking into an inanimate object was possible. After all, from a conceptual standpoint, AI is the assumption that the processing of human thought can be mechanized. Multiple civilizations including the Chinese[16] & Muslims[17] have portrayed this through more or less developed automatons, which are self-operating machines.

Philosophers such as Aristotle, Euclid & AL-Khwarizmi all contributed to AI in their own way.[18] Aristotle defined syllogism, therefore, enabling logical arguments to be developed through deductive reasoning.  Euclid showcases his formal reasoning through a set of mathematical proofs & propositions, which are still taught today, in *Elements*[19]. AL-Khwarizmi found the first systematic solution of linear and quadratic equations.[20] Without him, we would not have algorithms, the essence behind AI.

These elements provided a structure for AI, however, a common language was needed to express and develop these ideas. Although Leibniz made an attempt to create a universal language with his *Characteristica universalis* in the 17th

---

[15] McCorduck, Pamela (2004), Machines Who Think (2nd ed.), Natick, MA: A. K. Peters, Ltd., ISBN 978-1-56881-205-2, OCLC 52197627

[16] Ibid 17

[17] Nick, Martin (2005), Al Jazari: The Ingenious 13th Century Muslim Mechanic, Al Shindagah, retrieved July 08 2019

[18] Berlinski, David (2000), The Advent of the Algorithm, Harcourt Books, ISBN 978-0-15-601391-8,

[19] Todhunter, I. (1933). Euclid elements: Books I-VI, XI and XII. Dutton, NY: Everymans Library.

[20] Rāshid, R. (2009). Al-Khwārizmī: The beginnings of algebra. London: Saqi.

century, this did not work.[21] However, in 1913, *Principia Mathematica* by Russel & Whitehead showcased the logical validity of mathematics.[22] Now, the problem was not "Which language do we use?" but "Can this language say everything?" or in other words, "What are its limits?"

Alan Turing, a British savant many consider to be the father of AI, created the Turing Machine. Although, it isn't practically useful, this machine enabled the invention of programmable computers that can execute multiple tasks with symbols as simple as 1s & 0s. The fact that one computer could carry many different tasks was novel for the time. This innovation helped the Allies decrypt secret German codes generated by the famous Enigma machine during WWII.[23] This was estimated to have saved millions of lives.[24]

Today, no computational program can do more than a Turing machine. As a matter of fact, the programming languages used today such as C++, Java, Cobalt... are all said to be "Turing complete" meaning that they have the same computational power than the Turing machine.

Another computational model was developed in the 1930's at the same time than the Turing Machine, it was the Lambda calculus by mathematician Alonzo Church[25] that was also said to be Turing complete. These discoveries led to the fact that a mechanical device could emulate a mathematician and carry out logical deduction tasks within the limits set by Gödel's incompleteness theorems, which state that the axioms (the assumptions set in a system) are considered to be true but cannot be proved by the same system (explained in *To Go Further*). The creation of such machines implicitly suggested that the creation of an electronic brain was possible.[26]

---

[21] Buchanan, Bruce G. (Winter 2005), "A (Very) Brief History of Artificial Intelligence" (PDF), AI Magazine, pp. 53–60, archived from the original (PDF) on 26 December 2007, retrieved July 08 2019

[22] Ivor Grattan-Guinness (2000) The Search for Mathematical Roots 1870–1940, Princeton University Press, Princeton N.J., ISBN 0-691-05857-1

[23] Copeland, P. J. (2012, June 19). Alan Turing: The codebreaker who saved 'millions of lives'. Retrieved July 12, 2019, from https://www.bbc.com/news/technology-18419691

[24] Ibid

[25] Church, A. (1932). "A set of postulates for the foundation of logic". Annals of Mathematics. Series 2. 33 (2): 346–366

[26] Crevier, Daniel (1993), AI: The Tumultuous Search for Artificial Intelligence, New York, NY: BasicBooks, ISBN 0-465-02997-3

During the 50's, the first Neural Networks were created[27] and the famous Turing Test saw the day. The test measured the ability of a machine to 'think'. It used a pass or fail model stating that the machine was "thinking" if it could carry on a conversation over a teleprinter with a human without them being able to tell the difference. It is only in the mid 50's when researchers had access to digital computers that AI started to take off. In 1955, the Logic Theorist was created by Newell & Simon. It was the first computer program engineered to solve problems. As a matter of fact the program managed to prove close to 75% of the first theorems of Russel & Whitehead's *Principia Mathematica*[28]. Although the 'Logic Theorist' is considered to be the first AI program ever, it is only in 1956 that the term 'Artificial Intelligence' becomes the official name for this field of study[29] at the Dartmouth Conference organized by Marvin Minsky, John McCarthy, Claude Shannon & Nathan Rochester.[30]

This sparked the golden years of AI. MIT, Stanford and Carnegie Mellon University were heavily subsidized by the current Defense Advanced Research Project Agency (DARPA). Do not forget that most new technologies during this period were funded by the American government's efforts to outdo the Russians during the Cold War. **Technological innovation is never innocent, there is always something more than just a scientific motive.**

Overseas, Edinburgh received funding by the British government. These institutions led AI research for years to come.[31] Minsky & McCarthy, profiting from their image as the founders of AI from 1956, attracted funding to their research programs in MIT and Stanford while Newell & Simon surfed on their previous success from the 'Logic Theorist' at CMU. These golden years also led to the realization of certain projects such as 'ELIZA', the first chatterbot making it seem like you could communicate with it until you realized it just rephrased what it was told.[32] Also, 'WABOT-1', the first humanoid robot saw the day in

---

[27] McCorduk (n 14) 51-57

[28] Crevier, (n 23) 44-46

[29] Ibid 49

[30] Russell, Stuart J.; Norvig, Peter (2003), Artificial Intelligence: A Modern Approach (2nd ed.), Upper Saddle River, New Jersey: Prentice Hall, ISBN 0-13-790395-2 p.17

[31] Crevier, (n 23) 64-65

[32] McCorduck (n 14) 291-296

1972. Created by the Japanese, this android could walk, hold objects and communicate with a person.[33] The researchers in the field were optimistic due to the inflows of money and the promising progress being made, however, they had underestimated the time it would take to meet the sky-high expectations of their funders. Consequently, the funding of AI stopped in the 70s.[34]

The main reasons behind this crisis, according to Hans Moravec, was the limited computer power for tasks such as vision and natural language, which were the main focus of the funders, combined with the exaggerated predictions researchers made.[35]

In the 1980s, the focus shifted to knowledge. 'Expert systems' took the spotlight, which pushed the Japanese government to fund AI.[36] These systems emulated the decision-making of experts through conditional branching (if-then rules). This made it easier for non-experts to find the answers they are looking for in an efficient manner and marked the beginning of another run.

If we take the definition of AI given in chapter 1, 'expert systems' wouldn't be considered AI today. Nevertheless, remember, AI is relative, what would be considered AI one day can be simple statistics, probability or else ten years after.

In 1982, the concept of neural networks was back in the spotlight with the "Hopfield nets"[37] created by John Hopfield. This revived the field of connectionism after Minsky's & Papert's book *Perceptrons: an introduction to computational geometry* which criticized and underlined the limits of this concept brought forward by F. Rosenblatt in 1958. After its release, the field was buried and left untouched for years as researchers knew that their work would not attract funding. No one expected neural networks to be the way to go for AI until Hopfield entered in 1982 and sparked another boom.

---

[33] Ichbiah, D. (2005). Robots: From science fiction to technological revolution. New York: Harry N. Abrams. p. 130

[34] Russell & Norvig (n 27) 21-22

[35] Crevier (n 23) 115

[36] Feigenbaum, Edward A.; McCorduck, Pamela (1983), The Fifth Generation: Artificial Intelligence and Japan's Computer Challenge to the World, Michael Joseph, ISBN 978-0-7181-2401-4

[37] Russell & Norvig (n 27) 25

In 1987, the field lost its momentum when specialized AI hardware was caught up in terms of computing power by desktop computers made by Apple & IBM.[38] Researchers then shifted the focus to robotics citing the need for a body in which you need to put a mind while referring to AI.[39]

After this last "AI Winter", term which was given to these less glamorous periods of AI with low funding[40], the scene was being set from the late 80s to 2010 for AI to become what it is today.

AI used to be limited by computer power, not anymore. As expressed by Moore's law, the number of components processing data, such as transistors, doubles every two years.[41] With this additional computer power, more refined models could be implemented such as Bayesian networks, stochastic models such as Markovs'... (Detailed in *To Go further*).This enabled projects such as Deep Blue & Watson to see the day. Deep Blue, mentioned in Chapter 1, was the first computer to beat the reigning world chess champion, Garry Kasparov in 1997.[42] Watson was a question answering system created by IBM that put two *Jeopardy!* Champions to shame.[43] As you can see throughout time, AI is mostly used to solve 'games' (I put games in brackets because depending on the game played, the outcomes could vary. For example, Deep Blue could only damage the ego of a chess player while Turing's machine indirectly saved lives in WWII).

Although progress was being made and the application of AI tools had been successful at times, the sentiment on AI overall was negative.[44] Therefore, researchers categorized their project under other 'fancier' names for a few years such as informatics, cognitive systems, etc.

In 2009, the term of Big Data gained popularity thanks to a paper by the McKinsey Global Institute in which it is said that companies using Big Data will

---

[38] McCorduck (n 14) 435

[39] McCorduck (n 14) 454-462

[40] Crevier (n 23) 203

[41] Moore, Gordon E. (1965). "Cramming more components onto integrated circuits" (PDF). Electronics Magazine. p. 4. Retrieved 2019-11-07

[42] IBM (n 3)

[43] Markoff, J. (2011, February 16). Computer Wins on 'Jeopardy!' Trivial, It's Not. Retrieved July 11, 2019 https://www.nytimes.com/2011/02/17/science/17jeopardy-watson.html

[44] Tascarella, P. (2006, August 14). Robotics firms find fundraising struggle with venture capital shy. Retrieved July 11, 2019, from https://www.bizjournals.com/pittsburgh/stories/2006/08/14/focus3.html?b=1155528000^1329573

be able to allocate their resources better and be more productive overall[45] (Big Data simply refers to large volumes of data that normal programs cannot process). Consequently, this pushed machine learning upfront too as it is necessary to analyze Big Data.

Since then, AI and technology overall have been on a crazy train. Why? Whether it's the song that'll make you happier or the analytics a business can exploit to increase its sales or the diagnostic a doctor can make to save a life, data is knowledge and knowledge is a powerful resource. Unlike other commodities, data is infinite, it's intangible and can be created instantly and is extremely easy to store. Therefore, it can grow exponentially. "AI is the new electricity" according to computer scientist, Andrew NG[46] due to its potential to be omnipresent and disrupt the way we live, work and interact with each other. AI can already be seen in our everyday lives.

If you use Facebook, Instagram or any type of social media, your preferences, likes and all the content you gravitate towards within those applications is being processed by AI tools, which then push back personalized content that you may like.[47] Spotify and Netflix also operate in the same manner, they keep track of what you listen and watch and recommend certain types of content.[48] To go further in the personalization example, let's have a look at Google. Did you ever wonder why ads for the exact objects or services you need, such as a brand-new TV or a holiday package, show up while you're browsing the web? Well, an AI tool figured what you wanted (or is trying to figure what you want) based on a recent query you made or through deduction work with inferential statistics. For example, let's imagine you have been looking up symptoms of burnout or depression and have been searching for ways to increase your energy level. Google's AI systems will show you ads of holidays because many

---

[45] Manyika, James; Chui, Michael; Bughin, Jaques; Brown, Brad; Dobbs, Richard; Roxburgh, Charles; Byers, Angela Hung (May 2011). "Big Data: The next frontier for innovation, competition, and productivity". McKinsey Global Institute. Retrieved July 11 2019.

[46] Lynch, S. (2017, March 11). Andrew Ng: Why AI Is the New Electricity. Retrieved July 12, 2019, from https://www.gsb.stanford.edu/insights/andrew-ng-why-ai-new-electricity

[47] Safian (n 6)

[48] Ibid

people sharing the same demographics and behaviors online as you have taken holidays. People that believe in fate might just be fooled by AI in the future...

For everything to be personalized, these tech companies must know your demographics, preferences, dislikes and overall behavior. Why do they want to find out? The companies that can track this data better can monetize it. Whether it's by enhancing your experience and charging you more or simply making it easier for businesses to target you as a potential customer, it always comes down to the bottom line.

Let's say you do not use computers or social media. Can you escape AI?

If you drive or use roads, you will always be interacting with AI in the upcoming years one way or another whether you are in self-driving vehicle or being assessed by one. If you like movies or come across any visual content, most of the software systems used to edit images and add special effects use AI to a certain extent. For example, Josh Brolin's facial expressions were captured and fed to an AI algorithm to ensure the accurate fitting onto Thanos' face in the Avengers movies.[49] AI is also being used to evaluate the commercial appeal of screenplays and some believe that one day, AI systems will be able to write and produce their own movies[50], therefore, completely disrupting the sector (this last part will be developed in Chapter 3 due to the challenges this represents).

In a nutshell, no you won't be able to escape AI, especially when you think about the security sector. Certain private companies already use AI to identify people with facial recognition. This was shown to be successful during a concert by Jacky Cheung, a Hong Kong pop star.[51] More than 12 criminals thought they could enjoy a concert in peace and never get caught within the masses, however, they were quickly identified and arrested thanks to the AI-assisted CCTV system.[52] The applications for Defense are endless. We have already seen that

[49] Robitzski, D. (2018, July 18). Artificial intelligence is automating Hollywood. Now, art can thrive. Retrieved July 15, 2019, from https://futurism.com/artificial-intelligence-automating-hollywood-art

[50] Ibid

[51] Lee, K., Dr., & Teddy-Ang, S., Dr. (n.d.). AI Super Powers: China, Silicon Valley and the New World Order. Lecture. Organized by the Lee Kuan Yew School of Public Policy

[52] 'Crime-fighter' Jacky Cheung adds 12 crooks, 2 drones to his tally. (2018, September 24). Retrieved July 15, 2019, from https://www.scmp.com/news/china/society/article/2165566/chinas-crime-fighting-pop-star-jacky-cheung-adds-12-crooks-two

wars can be fought at a distance with drones flying over conflict zones with its pilot commanding it in a safe zone thousands of miles away. Imagine if they could track and eliminate threats with AI. Project Maven almost makes this true. It is a ML-based system that scans drone video footage to find potential targets (individuals, buildings, vehicles) to bomb.[53] As mentioned before, AI and war are extremely close since the mid-20th century. The country that will win this "algorithmic warfare" has a huge edge against the others. Luckily, this is not the preferred use governments will make of AI.

Governmental bodies have already implemented and will continue to implement AI systems to help, protect and monitor their citizens. These systems will assist the public sector employees track and allocate social benefits, detect any fraudulent claims...[54] Governments will have the capacity to track their citizen's behaviors.[55] Everyone is prone to be exposed to AI. Living completely off the grid will be extremely difficult in the future. Is this beneficial due to the level of information countries and businesses will have to help us or detrimental due to the fact that we won't even be knowing who owns our information and who can be potentially stealing our individual data?

---

[53] Tarnoff, B. (2018, October 11). Weaponised AI is coming. Are algorithmic forever wars our future? Ben Tarnoff. Retrieved July 15, 2019, from https://www.theguardian.com/commentisfree/2018/oct/11/war-jedi-algorithmic-warfare-us-military

[54] Institute of Public Administration Australia. "In Brief - Artificial Intelligence in the Public Sector". Linked infographic based on information by Daniel Castro, Steve Nichols, Eric Ellis, Cynthia Stoddard (Adobe Chief Information Officer) and Government Technology reporting. Retrieved 2019-07-15.

[55] CBS News. (2018, April 24). China's behavior monitoring system bars some from travel, purchasing property. Retrieved July 15, 2019, from https://www.cbsnews.com/news/china-social-credit-system-surveillance-cameras/

# Chapter 3:    The Impact of AI

Is AI good or bad? Well, it depends… This was probably not the answer you are looking for. AI systems can be beneficial in endless scenarios as seen in the previous examples. From receiving customized services that suit your needs, automatizing repetitive tasks to stopping criminals and increasing productivity. AI sounds great at first, nevertheless, in the end, it all depends on who get their hands on your data.

Let's take a closer look at how these AI-using entities get their data. They spy on you… with your permission. For example, Amazon's Alexa is an AI based personal assistant that is voice activated. If you read the terms of use, section *1.1 General* says "Amazon processes and retains your Alexa Interactions and related information in the cloud…"[56] and section *1.3 Contacts* "Amazon will periodically import and store your contacts to improve your Alexa Communication experience."[57] As soon as you use the product, you are considered to have accepted the terms, whether you have read them or not. The reason I used the word "spy" is because of the many people that do not realize that they are giving up control on their private life. They are seduced into using the latest tech gadget, agree to the never ending paragraphs of fine print (if there are any) and forget that they have invited Amazon into their home. Depending on how often you'd use Alexa, Amazon can eavesdrop into any conversation you're having, whether, it's personal or business-related. It'll know through sound how many people are in your house at all time and who

---

[56]   Alexa   Terms   of   Use.   (n.d.).   Retrieved   July   16,   2019,   from https://www.amazon.com/gp/help/customer/display.html?nodeId=201809740
[57] Ibid

these people can be when relating it to your network and text messages. Alexa will have enough information to identify whether you're having an affair or not! Did you realize this? If yes, would you still buy this product? Is having an AI virtual assistant help you schedule your meetings and play your music worth giving up your whole private life? Amazon's Alexa is one of the many products and services tech companies use to have an in depth look at you. In addition to Amazon, Microsoft, Apple, Alphabet Inc. (Google), Facebook, Alibaba and Tencent are amongst these Ai-using, tech heavyweights that have already so much data on individuals through various products and services that they probably know these people better than their own countries, friends, partners and even themselves. This is one of the factors that the 7 companies mentioned above are part of the top 10 in terms of market capitalization in the world.[58]

## Top 10 Global Market Capitalizations as of 28.06.19

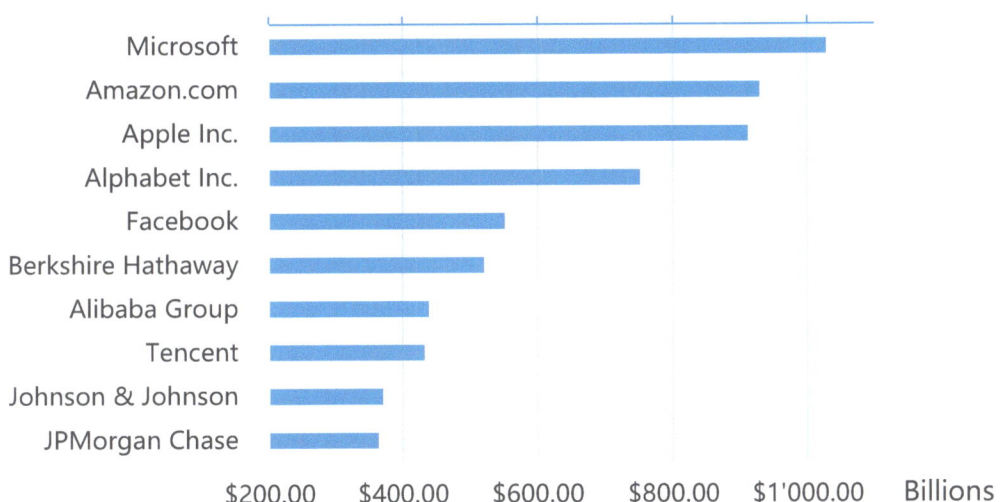

Source: Bloomberg[58]

Although, these market capitalizations seem excessive as some are flirting with the trillion dollar mark and are larger than most countries GDPs[59], they are also

---

[58] Source: Bloomberg as of 28.06.19 (Retrieved 16.07.19)
[59] Companies that have more money than entire countries. (2018, July 28). Retrieved July 16, 2019, from https://www.rt.com/business/434516-companies-countries-gdp-revenue/

based on the potential power these companies have due to their access to data and their ability to interpret it. If one of these tech companies wanted to overthrow a government, they probably could. Of course, this probably will not happen as it is not the wisest action for a tech giant to take. However, the Cambridge Analytica scandal exposed Facebook's reach and the impact their data could have when put into the wrong hands[60]. Data has the potential to be a modern weapon of mass destruction as it can influence people and seal governments' fates. The question now is who has this potential weapon?

The majority of the companies mentioned before come from two places: the USA and China. They are the two drivers of today's global financial markets and happen to be in a technological war. This new tech cold war was sparked by US president, Donald Trump, who wanted to promote "Made in America" and limit Chinese "trade abuses" in 2014.[61] Effectively, certain American businesses were behind as they could not price their products as cheap as their Chinese counterparts due to labor costs. Trump leveled the field with a series of tariffs imposed on certain products to which Chinese president, Xi Jinping, retaliated with his own set of tariffs on American products. This trade war has been going on since 2016.[62] An interesting case of this technological cold war is Huawei. The communications company became the 2nd best phone seller in the world in 2018 with cheaper iPhone-like devices. In February 2018, FBI director Chris Wray stated "We're deeply concerned about the risks of allowing a company or entity that is beholden to foreign governments that don't share our values to gain positions of power inside our telecommunications networks [...] (that power would give them) the capacity to exert pressure or control over our telecommunications infrastructure, it provides the capacity to maliciously modify or steals information and provides the capacity to conduct undetected

[60] Graham-Harrison, E., & Cadwalladr, C. (2018, March 17). Revealed: 50 million Facebook profiles harvested for Cambridge Analytica in major data breach. Retrieved July 16, 2019, from https://www.theguardian.com/news/2018/mar/17/cambridge-analytica-facebook-influence-us-election

[61] Jacob Pramuk, J. W. (2019, June 29). US and China agree to continue tariff talks. Here's a timeline of how the trade war started. Retrieved July 16, 2019, from https://www.cnbc.com/2019/06/29/us-china-trade-talks-at-g-20-timeline-of-how-the-tariff-war-started.html

[62] Wong, D. (2019, July 03). The US-China Trade War: A Timeline. Retrieved July 16, 2019, from https://www.china-briefing.com/news/the-us-china-trade-war-a-timeline/

espionage."[63] After this statement, American carriers and distributors such as AT&T and Best Buy started dropping Huawei.[64] The Pentagon banned the Chinese phones on military bases and any remaining US company that dealt with them was scrutinized by the Trump administration. Google and Facebook were both called out by the Congress for their ties with the Chinese brand during summer 2018.[65] America's soft power could also be witnessed as Japan, Australia and Canada followed and took measures against Huawei. Canada even arrested Huawei's CFO, Meng Wanzhou on US request for supposedly skirting American sanctions on Iran[66] at the end of 2018 when Huawei's sales exceeded 200 million smartphones. Later, Trump banned Huawei on May 15th creating a $30 Billion revenue loss for the Chinese company and restricted US companies to sell their components to Huawei.[67] As of today[68], USA & China are still going back and forth.

I am not sure whether Huawei had the intention to spy on their users or if they were just sabotaged by president Trump because of their stellar growth and their threat to the US in the race to 5G.

Another introductory book could be written about the technology behind 5G. To make it short, 5G is the newest generation of mobile networks with the largest wireless bandwidth ever seen.[69] It's about 20 times faster than our current 4G network.[70] This would enable machines, robots and other devices such as self-driving cars to interconnect.[71] Imagine vehicles that would never

[63] Boom, D. V. (2018, February 15). Don't use phones from Huawei or ZTE, FBI director says. Retrieved July 16, 2019, from https://www.cnet.com/news/huawei-zte-fbi-chris-wray-nsa/

[64] Keane, S. (2019, July 17). Huawei ban: Full timeline on how and why its phones are under fire. Retrieved July 17, 2019, from https://www.cnet.com/news/huawei-ban-full-timeline-on-how-and-why-its-phones-are-under-fire/

[65] Ibid

[66] Kharpal, A. (2019, May 08). Huawei CFO's extradition case: Everything you need to know. Retrieved July 19, 2019, from https://www.cnbc.com/2019/05/08/huawei-cfo-meng-wanzhou-extradition-case-everything-you-need-to-know.html

[67] Keane (n 63)

[68] Written on July 19th 2019

[69] What is 5G?: Everything You Need to Know About 5G. (2019, April 12). Retrieved July 19, 2019, from https://www.qualcomm.com/invention/5g/what-is-5g

[70] Fisher, T. (2019, July 03). 5G vs 4G: Everything You Need to Know. Retrieved July 19, 2019, from https://www.lifewire.com/5g-vs-4g-4156322

[71] Qualcomm (n 67)

crash because they can communicate with each other. Therefore, 5G represents an important extension to any AI system and will be a catalyst for the development of AI tools.

Coming back to our tech war, the leader in 5G equipment right now is Huawei. They happen to be a private Chinese company but are believed to have ties with the Chinese government. The USA, amongst other countries, fear that the Chinese are attempting to spy and gather data through Huawei. This is a valid concern since most of the 5G equipment is currently being rolled out by Huawei in multiple countries. It would be extremely scary if the Chinese were to spy and steal data from the countries where they have installed their equipment. China could simply rule the world with all the information they would have. They would be able to predict every single move of their opponents in case of a conflict or could simply know what everyone is doing. The 5G battle has been won by China without a doubt but who's winning this tech war?

If we look at our market capitalizations chart, Americans companies dominate. Out of the 7 top companies 5 are American and only 2 are Chinese.

In terms of users, China has the advantage as it has more depth with 1.5 Billion wireless users which represents more than 3 times the number of American users.[72] From a data acquisition & analysis standpoint, this is a huge advantage to test and improve AI systems pertaining to human behavior. Remember, in the AI world, the more information you have the better. Although the Chinese have more users, they do not have the same spending power as the Americans, the American GDP per capita sits around $60k in 2018 while the Chinese is below $20k.[73] This gives Chinese companies less freedom to price their products as high as the Americans.

From another angle, talent follows money, therefore, talents are also attracted to the USA due to the better quality of life they can have. The USA has 56% more talent than China does[74]. With more talent and wealthier clients, American

---

[72] Wu, Debby, et al. "Who's Winning the Tech Cold War? A China vs. U.S. Scoreboard." Bloomberg.com, Bloomberg, 20 July 2019, www.bloomberg.com/graphics/2019-us-china-who-is-winning-the-tech-war/.
[73] Ibid
[74] Ibid

companies naturally have a bigger market and can book higher revenue than their Chinese counterparts. This quickly snowballs. With higher revenue, a positive feedback look sets itself as companies can invest more in R&D which then enables them to have better products, they can then price those products higher and this creates more revenue and so on.

In conclusion, this tech war is close. USA is leading, however, the potential of the Chinese to overthrow them is at an all-time high. Without Trump's aggressive moves, this may have already been the case. The depth of their data, their access to 5G and the rise of companies such as Huawei exemplify why the gap between the two superpowers is being closed. The only missing element is that China doesn't innovate yet, they are great at copying and can outdo the US through their numbers. Nevertheless, they lack a little spark to attract talent. They do not have the soft power the USA has. For example, what would you choose between relaxing on sunny Californian beaches and dipping your toes in the South Chinese sea, which may or may not been polluted by Japanese nuclear waste?
Opinions might differ but I don't know many people who say that China is their #1 global destination.

Until now, we have seen what AI can do and its effects on global economies. What about its social effects? How does AI affect us on an individual scale? Other than affecting our privacy, AI will change our behaviors. New technologies such as mobile phones and the internet have changed our lifestyles tremendously compared to 20 years ago. From a personal standpoint, I have seen our physical social interactions diminish. A few years ago, people spoke to each other in buses, bars and the outdoors in general. Now, the younger people are always on their phone or have their music plugged in. They are maybe a few meters away or in the next room and they would rather text than get up and speak to the person. Even I am guilty of texting my coworkers on our office chat when they are 10 feet away! Anyone that has worked in an office that has gone through a major computer update or migration knows how messy these transitions can be. Some of us get lost after a simple email server update. Your environment is disturbed, elements have changed places and you need to learn everything again from the beginning. For the younger users, this is just a matter of making a few clicks. For the elder ones, it is more complicated...If these simple updates can cause quite a stir, imagine transitioning to a world with AI systems! The key factor differentiating between the success or complete disaster of AI is the way we transition into it.

As we transition into this AI world, we can identify different phases:

-An initial phase, where AI is slowly being implemented but most of the work is done by humans. This is the one which we are in today.

-A transition phase, where humans & AI will share similar tasks as the AI systems wouldn't have been fully implemented where they could be.

-A final phase, where AI would have been completely implemented in places where it would make sense to have it (this will become clearer in the next pages)

The start will be slow. For the smaller systems, you'll most likely have to enter plenty of parameters to set the AI system in the right way. Amazon's Alexa for example must ask you a dozen questions before you start using it. With time, these ML systems and the companies creating them will gather more data in order to ameliorate the user experience and make the setting up of their products and services easier.

On a larger scale, in robotics for example, the transition phase where human workers will be mixed with robots can quickly become dangerous. If humans interact with humans, each person is responsible for their actions. This corresponds to the situation today. For example, most cars today are controlled by humans, everyone is aware of their surroundings when they drive since they consciously drive their car. When AI will completely take over this field with self-driving cars, the systems will know where the other cars are and will communicate with each other to avoid collisions, which makes the environment much safer than it is today. The tricky part once again is the transition. How do we go from today's world with a majority of human drivers to tomorrow's with only self-driving cars? There is no choice other than having human drivers on the roads with self-driving cars. Humans do not have any boundaries dictated by an algorithm unlike AI systems, they are unpredictable. To solve this, the autonomous cars must be more precautious with additional safety measures to take into account the human factor. This is what Tesla and other self-driving cars are doing. However, if this is badly done and leads to an accident, AI will be at fault today because it is something relatively new. If two human drivers have an accident, no one cares, it happens every day. If a self-driving car is involved, it will make news even if it took the safest action possible. To take the example further, if a self-driving car knows an accident is inevitable, what should it do? After all they are governed by a set of advanced algorithms. For the sake of the argument, let's say the car is going down a narrow road, there's a wall on one side, a huge tree on the other and a young girl jumps in the middle of the road and the car does not have the time to break. Does the car run into the wall or the tree to avoid an inevitable fatal hit because you have a better chance of survival as a driver or does it go straight and does it best to brake to hit the girl as slow as possible in order to keep the driver safe? Today, there are no written laws about this, but computers are making their decisions based on the data they are being fed. As more technology is being developed and implemented, these rare occurrences will occur, and lawmakers & regulators will have to intervene. This will not be exclusive to self-driving cars, it will also apply to other systems across sectors: medical, legal, financial, educational, agricultural.... The issue with this is that regulations made for the transition phase where humans & AI interact may not fit the final phase. Coming back to our previous example, let's say the car ran over the girl. The judge decided it was the car's fault (who will take responsibility with AI systems is another never-ending debate that I will not address in this book) and therefore, reduces speed limits in similar areas to avoid future accidents. This works greatly during the transition phase. Now, by the time we get to the final phase, roads are safer, sensors and brakes are better, the speed limit is not needed anymore as the car would be able to stop in time. Depending on the reduction of the speed limit, this will negatively affect the efficiency of the traffic because of a new law that has become irrelevant. In conclusion, laws will need to be made accordingly and readjusted with time. An ML system will probably in charge of doing this.

The previous example of self-driving cars looks perfect in its final stage where every car is aware of its peers and since they all have similar decision-making, the environment should be predictable and therefore, safe. Let's take another example where the implementation of AI in its final stage may not be as successful. In sectors where AI systems have to compete against each other like robot financial managers, the chances that they make the same decisions are extremely high since they would all be learning from the same data sets. The ML algorithm will go on and on until the market stays flat and all these robots have the same performance. In this case, the difference maker will be human. All in all, today some sectors look like they need AI, however, once the whole sector will have it, you need to bring back a human touch to it. The best businesses in the future will be the ones that manage to achieve the delicate balance between the mix of human and AI input.

The main issue with striking the perfect balance is that everyone is so similar yet so different. In the supermarket I go to, you find two types of cashiers, the physical ladies and the machines that scan your articles and check the weight of your basket to see whether you are stealing or have forgotten to scan anything. I go to the machines since it's faster and they are vacant most of the time because only a few people understand how they work. Others love to interact with the cashier including a friend of mine whose name I won't mention. He loves to buy an item that he doesn't need just to flirt with the most attractive cashier he can find. Needless to say, he will definitely be disappointed if machines replace all the young lady cashiers. Removing humans from certain types of jobs removes some charm from your daily routine… On the other hand, some of my other friends go with the human cashiers simply because they are lazy and love to have someone do everything for them, even if it takes longer and is less efficient.

This underlines another source of friction from the transition phase. AI can also make us lose our jobs. Kai Fu Lee did great work on this subject describing what type of jobs AI would affect in his whitepaper available online *Job Displacement Index*.[75] To summarize his work, the former Vice President of Google explains that jobs that are complex & strategic, require creativity, compassion, human touch and/or the ability to learn new environments are safe. However, any job that consists in repetitive tasks in a fixed environment or that doesn't require a specific skill will be replaced by AI.

Let's go through some examples. If you're an entrepreneur, CEO or have any leading, strategic role that requires to innovate and make decisions in unique environments, you are safe. Psychologists, medical practitioners do not need to worry either as their tasks are complex and require human interaction. Same for artists as they need creativity, although, some AI systems have been able to produce digital images that have been auctioned at Sotheby's.[76] Of course, these are digital images that have been artificially recreated thanks to a two-part ML system. One to figure out patterns from existing paintings and generate an image based on the given data and another one, a discriminator, to find all the differences between what has been generated and real paintings. The system keeps producing images until the discriminator finds no difference between the AI generated image and the real thing. Although the results are good, this is not an exact science. Art, movies, music and the other creative mediums will never be mastered by any AI because AI can only copy and the best artists steal.

Additionally, Kai Fu Lee also explains that jobs requiring dexterity will be safe as robots today aren't adroit and smooth as required. Since software progresses faster than robotics, Lee is right today, however, I believe that even jobs

---

[75] Accessible at https://aisuperpowers.com/job-displacement
[76] Vincent, James. "A Never-Ending Stream of AI Art Goes up for Auction." *The Verge*, The Verge, 5 Mar. 2019, www.theverge.com/2019/3/5/18251267/ai-art-gans-mario-klingemann-auction-sothebys-technology.

requiring dexterity will be in danger on the long term. So what jobs are in immediate danger? If you are a driver, a cleaner, a simple worker in a fast food restaurant or retail store, a controller, a telephone operator or have a job that consists in simple repetitive tasks, I have bad news for you... You can find the exact probability of your jobs disappearing in Kai Fu Lee's previously mentioned index.[77]

A few people have a social life exclusively because of their jobs. They have extremely demanding schedules that make them wake up early and go to bed late, making their coworkers their only social interaction for the day. What if AI replaced some of their coworkers? The mood would be different. Human interaction is a necessity for us to feel good. The "love hormone": oxytocin, which is a hormone that makes us happy by triggering the release of another hormone called serotonin, is produced during human interaction, physical touch, etc....[78] Simply speaking a few words with a person you appreciate can result in the triggering of the hormone. This explains why my friend stays in line for hours to speak to the cashier instead of going through the automated cashier.

If your space is filled by AI robots or empty because all the work is done by an AI software, I believe that you would not feel as fulfilled and will probably drown in depression if you are part of the "lucky" ones that kept their job. Some would argue that the gain in productivity with AI tools would give these employees the opportunity to spend more time outside of work, which would then result in more interactions with their friends. However, humans are creatures of habit. If they have been working 10 years straight on a slave-like rhythm and you suddenly yank them off, it won't be sunshine and blue skies for everyone... Some will be depressed and show withdrawal symptoms like a drug-addict.

This maybe a far-fetched example since we have not progressed to this stage, but it is another way to underline the importance of the speed we transition

---

[77] Lee (n 75)
[78] Bergland, C. (2013, September 12). The "Love Hormone" Drives Human Urge for Social Connection. Retrieved July 22, 2019, from https://www.psychologytoday.com/us/blog/the-athletes-way/201309/the-love-hormone-drives-human-urge-social-connection

into an AI world. If robots replaced your co-workers from one day to the other, this would be brutal, and you'd notice it right away. If a few coworkers are laid off every year, the transition would be smoother and would enable you to adapt.

Tying back to Lee's work, the jobs at risk are low skilled jobs meaning that anyone can access them without a high level of education. If we take a closer look at who is doing these type of jobs, we may see today clear warnings for the future.

Higher levels of education are correlated with higher family wealth.[79] Correlation is not causation, however, I will make the assumption that the people with those lower skilled jobs aren't the wealthiest people. If AI, which is pushed by the most capitalistic organizations led by the wealthiest part of the population, starts taking over the jobs of the poorer people, we are in for a bumpy ride! Inequality will increase furthermore, populism will rise[80] and we may just enter a social hellfire with civil wars or even revolutions. In my opinion, this is the biggest threats in our transition into an AI world. Ray Dalio and his Bridgewater associates observed the drastic rise of populism to an all-time high with their own index in 2017.[81] AI will inevitably be a catalyst to the ever-growing divide between the elite and the poorer. The implementation of AI systems can have terrible social consequences if the transition is made too fast without a safety net for the people at risk to lose their job. Either governments will have to create new types of "safe" jobs, slow down AI progress through regulation or they will have to educate their populations in order to work with and on AI. The last may be a blessing in disguise that would push the global population to a higher level of education. However, this seems to me as an ideal that will be difficult to strive towards as people are lazy as seen in the cashier example.

---

[79] Wolla, S. A., Dr., & Sullivan, J. (2017). Education, Income, and Wealth (Rep.). St. Louis, Missouri: Federal Reserve Bank of St. Louis. doi:research.stlouisfed.org Accessible at: https://files.stlouisfed.org/files/htdocs/publications/page1-econ/2017-01-03/education-income-and-wealth_SE.pdf

[80] Dalio, R., Kryger, S., Rogers, J., & Davis, G. (2017, March 22). Populism: The Phenomenon. Retrieved July 26, 2019, from https://www.bridgewater.com/resources/bwam032217.pdf

[81] Ibid

AI is a promising and disruptive technology that will greatly reduce the time taken to solve complex problems thus improving our productivity, increasing our safety and enabling us to focus on other areas to develop. Although we will always be far from what we see in the movies, AI has the capacity to have a huge social impact stimulating economies, shifting jobs and modifying the general behavior of populations. Rest assured, humans will always be needed to create, guide, interpret and adjust the AI systems which is the scariest part. AI cannot make disastrous decisions unless the person behind it is.

I hope you have enjoyed this introduction to AI, have learnt a few things in the field and were able to gain some insight on the social, economic and political implications AI has. Feel free to read the next chapter *To Go Further* which touches slightly more technical concepts that may not be accessible to readers without a basic knowledge in mathematics and statistics.

# Chapter 4: To Go Further

AI is an extremely wide subject as you have seen. Any sector can use AI in one way or another. Although I believe that AI goes way beyond the STEM field, algorithms are at the core of its mechanics. Therefore, coding is necessary to create your own AI program, which means that you will need a logical process to create an efficient one. Since there is no better language to express logic than mathematics, I have briefly compiled a few of the most common mathematical, statistical and probabilistic concepts you might encounter in the AI field.

I apologize in advance to all my teachers for attempting to explain in a few lines what they taught me over a few months.

## Regression Analysis

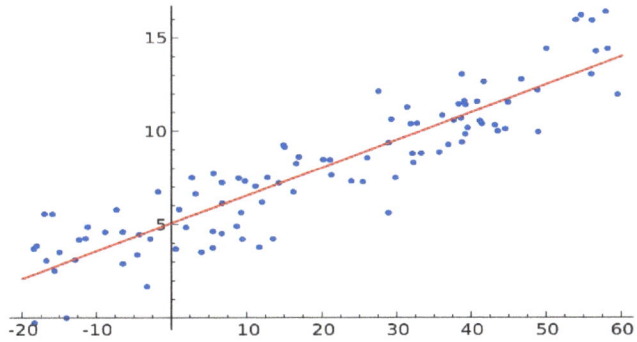

This statistical process estimates the relationships among variables. Linear regression will focus on the relationship between a dependent variable and an independent one. This is called multiple linear regression when there is more than one independent variable. In a nutshell, this finds the typical value of the dependent variable while the independent variable(s) change. For example, let's say you are studying the relation between the amount of water you drink and the temperature outside. Your x axis on the graph corresponds to the temperature you re in and the y axis to the amount of water you drink. You will gather different data points (in blue) and try to understand the relationship between the two. Regression analysis in this case draws a line of best fit to help you understand the relationship between both variables with the following formula:

$$y = a + xb$$

Where:          y is your dependent variable
                a is your y intercept
                x is your independent variable
                b is your slope

The above would be considered linear regression but could easily be changed to multiple linear regression (below) with multiple independent variables as long as they are non-collinear, meaning that they should not be highly correlated to each other. If this was the case, we would not be able to understand the distinctive relationships between each variable.

$$y = a + bX_0 + cX_1 + dX_2 + e$$

Where:          $X_0, X_1, X_2$ are your independent variables
                b, c, d are your slopes
                e is your residual error

Coming back to our example, if the above graph really depicted the relation between the amount of water you drink and the outside temperature, regression analysis would enable you to see that you drink more water when the temperature is higher, therefore, there is a positive correlation between higher temperatures and higher amounts of drank water. As you can see, if you were to this by hand for ten different independent variables, this can quickly become overwhelming. Therefore, ML systems are used to apply this simple process on Big Data, which can have millions of independent variables.

## K-means Clustering

In unsupervised learning, ML systems in need to have the capacity to filter data in different types of groups in order to mine data, which corresponds to recognizing patterns in large data sets. Whether it's for image recognition, identifying trends in stocks, behavior, etc. this is done by the K-means algorithm (also named Lloyd's algorithm) amongst others.

The K corresponds to the number of clusters you want to divide your data into. This algorithm can seem complicated, but it simply keeps calculating the distance between random points in the dataset until the variation between each cluster is at its minimum.

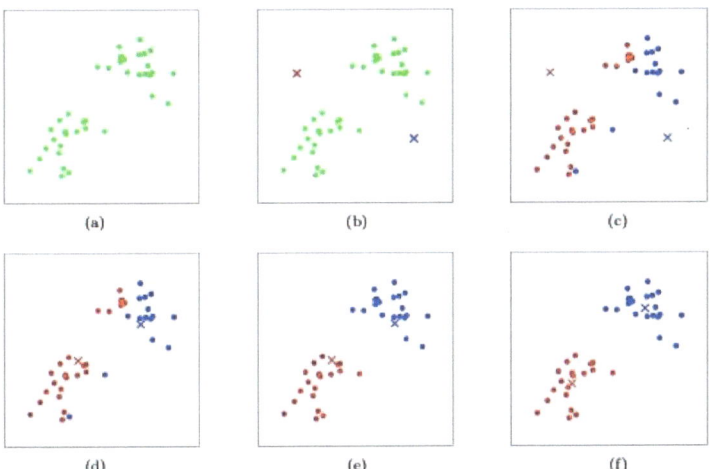

Briefly, you obtain your data set (a). You set k=2 as you want 2 clusters, therefore, you choose two random points (b). After calculating the distance between these chosen points and the others, initial clusters are formed (c). The process repeats itself with two other random points until the distance between each point of a cluster is minimized. If the total distance between the points of a cluster can be reduced, new clusters are formed (e). When the distance cannot be reduced anymore, the final clusters are set (f).

Mathematically, this can be translated to[82]:

1. Choose cluster centroids (fancy word to refer to the data point at the center of the cluster) $\mu_1, \mu_2, \mu_3 \ldots, \mu_k \in R^n$ randomly
2. Repeat until convergence:

   For every $i$, set $\qquad c^{(i)} := \arg\min \left\| x^{(i)} - \mu_j \right\|^2$

$$\text{For each } j, \text{ set } \mu_j = \frac{\sum_{i=1}^{m} 1\{c^{(i)}=j\}x^{(i)}}{\sum_{i=1}^{m} 1\{c^{(i)}=j\}}$$

Although this method is robust, it has limits such as choosing the number of clusters you want before you execute the algorithm, the need for the data to be linear and the inability to handle outliers or noisy data.[83]

### Decision Trees

These models use a multi-level conditional architecture. Decisions are made according to the preset rules. Each tree has a root node that splits into branches depending on the preset conditions to filter the original data. The number of branches and nodes between the root and lowest nodes can be infinite. However, the lowest layer nodes will not have branches as the decision tree would have come to an end as you can see below with the example of buying a car.[84] [85]

---

[82] Piech, C. (2013). K Means. Retrieved July 24, 2019, from https://stanford.edu/~cpiech/cs221/handouts/kmeans.html
[83] K-means clustering algorithm - Data Clustering Algorithms. (n.d.). Retrieved July 25, 2019, from https://sites.google.com/site/dataclusteringalgorithms/k-means-clustering-algorithm
[84] Rahul. (2018, August 08). Learn ML Algorithms by coding: Decision Trees. Retrieved July 26, 2019, from https://lethalbrains.com/learn-ml-algorithms-by-coding-decision-trees-439ac503c9a4
[85] Illustration from https://iq.opengenus.org

# Decision Tree for Classification and Prediction

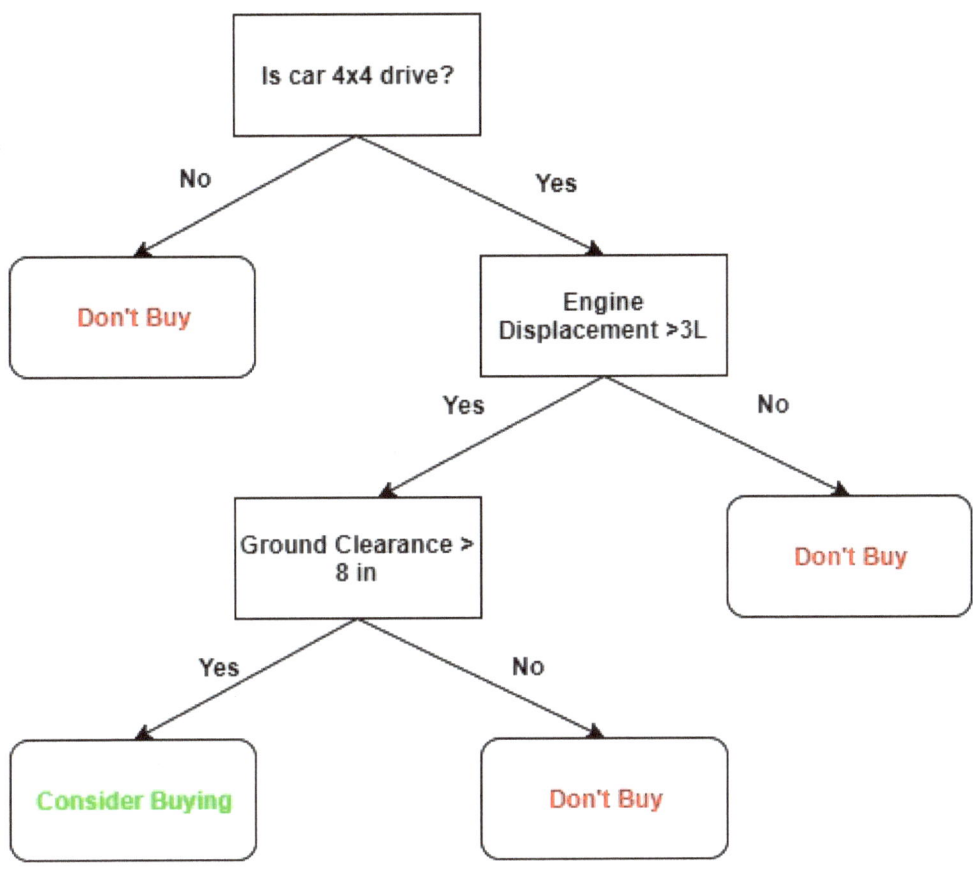

## Bayesian Model

This probabilistic model works best when trying to figure the likelihood of one event knowing that another event has occurred. For example, what is the probability that it rains Friday knowing that it has rained from Monday to Thursday?

Bayes Theorem states the following:

$$P(A|B) = \frac{P(B|A)P(A)}{P(B)}$$

Where :    $P$ is the symbol to denote probability.
$P(A|B)$ is the probability of event A occurring given that event B has occurred.
$P(B|A)$ is the probability of the event B occurring given that event A has occurred.
$P(A)$ is the probability of event A occurring
$P(B)$ is the probability of event B occurring

This theorem can be used by a decision tree. However, Bayesian models allow to solve more complex classification problems due to the sets of probability they render. Once a certain level of complexity is reached, decision trees would toss all information that would need to be separated down one branch. Finding where this error is made and correcting it can be tedious. On the other hand, a Bayesian model will give the different probabilities associated with each node to help its creator understand how it worked. Another advantage of this model is its ability to interpret networks with a directed acyclic graph[86] [87](below) unlike decision trees which end down their own branch and don't reconnect to future nodes down the tree.

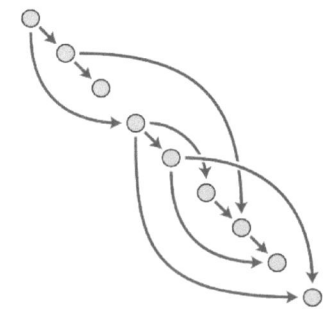

[86] Weisstein, Eric W. "Acyclic Digraph." From MathWorld--A Wolfram Web Resource. http://mathworld.wolfram.com/AcyclicDigraph.html
[87]Illustration by David Eppstein (2013)

To give you a better idea, spam filters can use this model to classify emails. For example, the system would classify an email as spam once it has reached a certain probability which would be affected by the sender and the  content of the mail, some trigger words increasing the probability of an email being spam could be "Nigeria" , "inheritance", "money"... etc.

## Inside Neural Networks

Remember the network from chapter 1? Let's take a deeper look inside the mechanism of these networks with information and algorithms from opendatascience.com.[88]

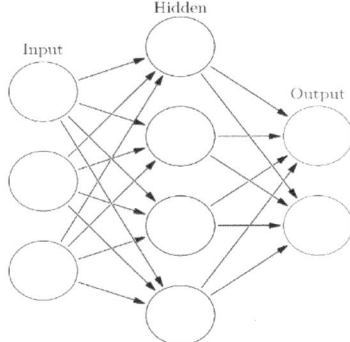

Let's take the example of an ML program that wants to predict the weather with a neural network. Relevant weather attributes such as time, humidity, air temperature, pressure, etc. will be the input values of our network. They would be fed forward to the hidden layer(s) until they reach the final output layer. While travelling through the network, each value is attributed a certain weight at each connection. Initially, weights are randomly picked. A feedforward algorithm is a form of guessing and looks like the following:

[88] Wylie, C. (2019, May 15). 5 Essential Neural Network Algorithms. Retrieved July 27, 2019, from https://opendatascience.com/essential-neural-network-algorithms/?source=post_page------------------------

$$n_l = S\left[\sum_{l-1}(w_l i_{l-1})\right]$$

Where:       $n_l$ is a neuron on layer $l$

$w_l$ is a weight value on layer $l$

$i$ is the value on the layer $l-1$

All input values are set as the first layer of neurons. Then, each neuron on the following layers takes the sum of all the neurons on the previous layer multiplied by the weights that connect them to the relevant neuron on that following layer.[89] This summed value is then activated. In order for these sums to be converted into a probability reflecting the weight of a neuron, they go through an activation algorithm that normalizes the output to a proportional value between 0 & 1. The most common one is the sigmoid function detailed below:

$$S(t) = \frac{1}{1 + e^{-t}}$$

Where:       $t$ is the value given by the feedforward algorithm

As non-linearity is introduced to the model, this network can give us more insight. However, since the weights have randomly been attributed, the output will be incorrect. Consequently, we have to train our ANN with huge datasets containing previous weather forecasts with the same attributes and the target values we are looking for so that we can compare the incorrect output to the desired target value. To compare the two, we need to know the amount of variation between the two, which corresponds to the level of error the system initially has. The squared cost function lets you find the error by calculating the difference between the output values and target values[90]:

---

[89] Ibid
[90] Ibid

$$E = \frac{1}{2}(n_l - T)^2$$

Where        $T$ is the target value

In order to make our network more accurate, we need to pass this error through a back propagation algorithm[91]. The output from the cost function is then multiplied by the derivative of the sigmoid function $S'$. Thus, $\delta$ is first defined at the output layer as the following:

$$\delta_L = \Delta E_{nl}(S'(n_L))$$

Then we need to calculate the error through each layer. Remember, error in the initial layers will transpose to the following ones. Therefore, any alteration in the past weight values must be transposed to fit the following layer of neurons.

$$\frac{\partial E}{\partial n} = \delta_L = [T(w_{l+1})(\delta_{L+1})(n_L(1 - n_L))]$$

This change can be traced back to an individual weight by multiplying it by the weight's activated input neuron value.

$$\frac{\partial E}{\partial w} = \frac{\partial E}{\partial n} n_{l-1}$$

The change now needs to be used to adapt the weight value and from this, we can figure out the learning rate of our system

---

[91] Ibid

$$w = w - (\eta \, \frac{\partial E}{\partial w})$$

Where:        $\eta$ is the learning rate

To summarize, the feedforward enables you to carry your information through the network, it randomly makes guesses so it will initially be incorrect unless you are extremely lucky. Therefore, you will need training data to backpropagate the network and adjust the weights by minimizing the error between your output and target values from your training data. Your neural network will become extremely precise after repeating this process a few thousand times for every piece of data especially, if you have different data sets you can train it with.

This in an introduction to get your own ANN started, there are much more powerful algorithms, both simple and complex that can be used to set up an ANN as the concepts above may have their limits in certain cases. Trivial aspects such as the data you feed the system, the level of bias the network (which can be reduced to a forced manual addition or subtraction on each layer) can have heavy effects on the output of these networks. If you have incorrect output, you can always correct it whether it's the data or the network in itself.

The most common issues you can encounter is the vanishing gradient problem while training your ML system. During the backpropagation step, each weight "receives an update that's proportional to the partial derivative of the error function with respect of the current weight in each iteration training"[92]. When using a sigmoid function for activation, all the values are between 0 & 1. Therefore, as you backpropagate, you multiply factors lower than 1 by each other making the gradient smaller and smaller. Consequently, the neurons in

---

[92] Goodfellow, I., Bengio, Y., & Courville, A. (2017). Deep learning. Cambridge, MA: MIT Press.

the earlier layers learn slowly and your prediction accuracy diminishes. To solve this problem, you can use other activation functions.[93]

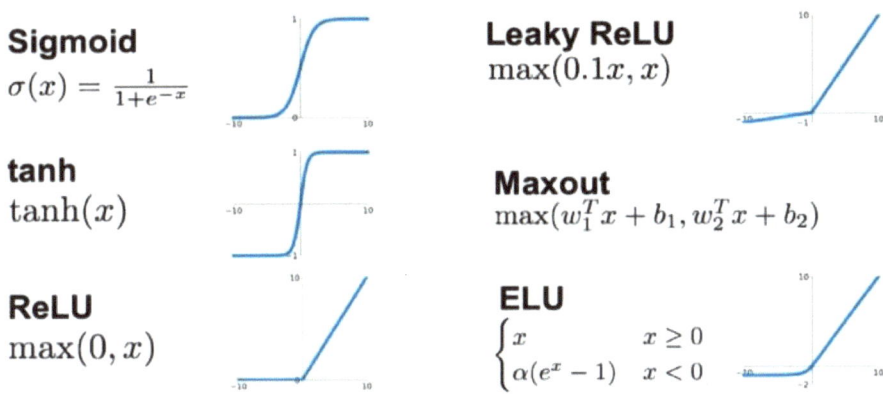

**Sigmoid**
$$\sigma(x) = \frac{1}{1+e^{-x}}$$

**tanh**
$$\tanh(x)$$

**ReLU**
$$\max(0, x)$$

**Leaky ReLU**
$$\max(0.1x, x)$$

**Maxout**
$$\max(w_1^T x + b_1, w_2^T x + b_2)$$

**ELU**
$$\begin{cases} x & x \geq 0 \\ \alpha(e^x - 1) & x < 0 \end{cases}$$

No matter the model you use, it always has to fit the data. If the model has too many parameters that cannot be justified by the data, the model will overfit the data.[94] On the other hand, if the model doesn't have enough parameters relative to the data, it will underfit it[95] as seen in the illustration below.[96]

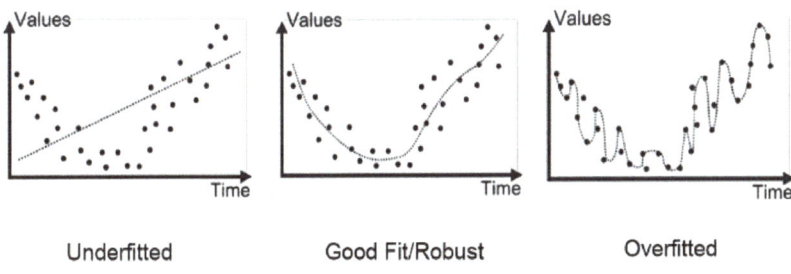

Underfitted          Good Fit/Robust          Overfitted

[93] Jadon, S. (2018, March 19). Introduction to Different Activation Functions for Deep Learning. Retrieved July 28, 2019, from https://medium.com/@shrutijadon10104776/survey-on-activation-functions-for-deep-learning-9689331ba092

[94] Information and Likelihood Theory: A Basis for Model Selection and Inference. (n.d.). Model Selection and Multimodel Inference, 49-97. doi:10.1007/978-0-387-22456-5_2

[95] Ibid

[96]Bande, A. (2018, March 18). What is underfitting and overfitting in machine learning and how to deal with it. Retrieved July 28, 2019, from https://medium.com/greyatom/what-is-underfitting-and-overfitting-in-machine-learning-and-how-to-deal-with-it-6803a989c76

In some rare cases, you might encounter an algorithm that keeps running without stopping, which is known as the halting problem.[97]

## Advanced Methods

Deep reinforcement learning is the most known method of Deep Learning. Google's Deepmind developed Alpha Go based on it.[98]

Deep Reinforcement Learning can be summarized as building an algorithm that learns directly from interaction with an environment. Like a human it learns from the consequences of its actions.

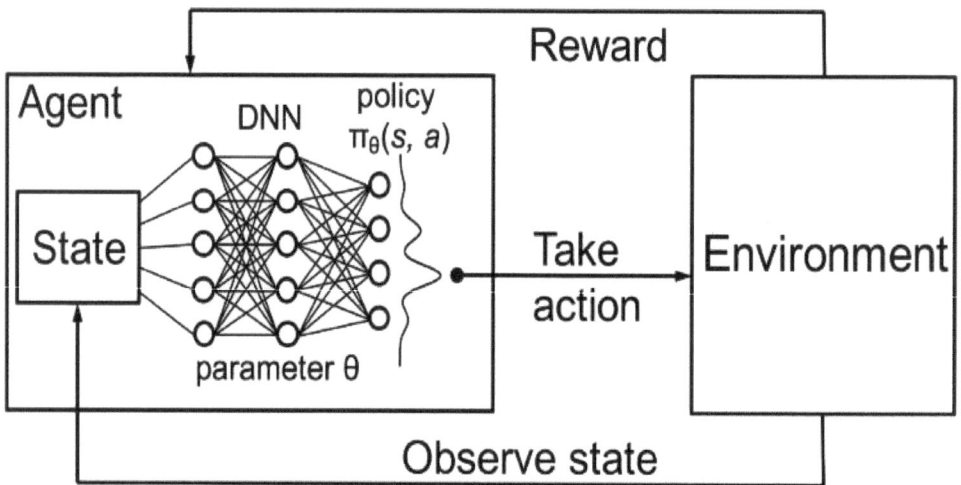

This method also uses a neural network. This network observes the current state of the environment and decides which action to take (e.g. move left, right etc.) in order to transition into another one. Based on the taken action the AI system receives a reward. The amount of the reward determines the quality of the taken Action with regards to solving the given problem (e.g. learning how to walk). After mentioning states, transitions and reward (referring to a utility

[97] Turing, A. M. (2016, December 23). London Mathematical Society Journals. Retrieved July 28, 2019, from https://londmathsoc.onlinelibrary.wiley.com/doi/abs/10.1112/plms/s2-42.1.230

[98] AlphaGo. (n.d.). Retrieved July 28, 2019, from https://deepmind.com/research/alphago/

function), math enthusiasts can discern that these systems might use some Markov processes which all revolve around one property: "the conditional probability distribution of future states of the process (conditional on both past and present states) depends only upon the present state, not on the sequence of events that preceded it."[99]

This mathematically translates to[100]:

$$\mathbb{P}[S_{t+1}|S_t] = \mathbb{P}[S_{t+1}|S_1, S_2, S_3, ..., S_t]$$

where      P represents a probability space

S represents a state

The transition from the current state $s$ to the next state $s'$ can only happen with a certain probability $Pss'$. In a Markov Process an agent that is told to go left would go left only with a certain probability of e.g. 0.998. With a small probability it is up to the environment to decide where the agent will end up.

$$\mathcal{P}_{ss'} = \mathbb{P}[S_{t+1} = s'|S_t = s]$$

$Pss'$ can be considered as an entry in a state transition matrix $P$ that defines transition probabilities from all states $s$ to all successor states $s'$.

$$\mathcal{P} = from \quad \begin{matrix} & to & \\ \begin{bmatrix} \mathcal{P}_{11} & \cdots & \mathcal{P}_{1n} \\ \vdots & & \\ \mathcal{P}_{n1} & \cdots & \mathcal{P}_{nn} \end{bmatrix} \end{matrix}$$

---

[99] Markov, A. A., & Schorr-Kon, J. J. (1968). Theory of algorithms. Jerusalem: Israel Program for Scientific Translations.

[100] The structure of Advanced Methods relies heavily on Artem Oppermann's work and uses his equations

The AI system will measure the efficiency of its decision-making with the following reward function. **R** is the reward that the system expects to receive in the state **s**.

$$\mathcal{R}_s = \mathbb{E}[R_{t+1}|S_t = s]$$

The most interesting is the total reward **Gt** which is the expected accumulated reward the system will receive across the sequence of all states. Every reward is weighted by so called discount factor $\gamma \in [0, 1]$. It is mathematically convenient to discount rewards since it avoids infinite returns in cyclic Markov processes. Besides the discount factor means the more we are in the future the less important the rewards become, because the future is often uncertain. If the reward is financial, immediate rewards may earn more interest than delayed rewards. Besides animal/human behavior shows preference for immediate reward.

$$G_t = R_{t+1} + R_{t+2} + \ldots = \sum_{k=0}^{\infty} \gamma^k R_{t+k+1}$$

For more information on the relations between Markov models and Deep Learning, I invite you to read the following series *Self Learning AI-Agents: Markov Decision Processes* by Artem Oppermann[101] who explains these in a simple but yet technical manner.

---

[101] Acccessible here: https://towardsdatascience.com/self-learning-ai-agents-part-i-markov-decision-processes-baf6b8fc4c5f

## The Future of the Future

Gordon Moore, former CEO of Intel, predicted that the number of transistors will double every two years in 1965. [102] Today, this still holds and is known as Moore's law. However, we are coming to saturation as the components cannot get any smaller as they come down to nanometers... or can they?

Although some of you may have been just introduced to AI and are still digesting the information contained in this handbook, you should be aware of quantum computing.

AI is the technology of the next decade, quantum computing is the technology of the one(s) after. Deep synergies will be formed between AI & quantum computing.[103] Maybe then robots will take over the world by accident... The best would be for you to do more research in your own time on this subject as it a confusing one.

Briefly, a quantum computer differs from a traditional computer as it is based on quantum physics and acts on particles such as electrons, photons, etc. A traditional computer stores information in a binary manner, a 0 or a 1, which is called a bit. In quantum computing, you have qBits, which do not have a binary form. Instead of having 0 and 1, a qBit will have a 20% chance of 0 & 80% of 1, or a 40-60 repartition amongst other combinations. This means that the qBit introduces uncertainty. This small tweak enables quantum computers to not only be more powerful than traditional computers but also completely different. They are believed to bring encryption to whole new level as hackers would have to break the laws of quantum physics to hack encryption codes. They might also be able to copy dynamic molecular structures, therefore, finding cures to illnesses. I have heard at certain forums that quantum

---

[102] Moore, G. (1998). Cramming More Components Onto Integrated Circuits. Proceedings of the IEEE, 86(1), 82-85. doi:10.1109/jproc.1998.658762

[103] Protalinski, E. (2019, March 15). ProBeat: AI and quantum computing continue to collide. Retrieved July 29, 2019, from https://venturebeat.com/2019/03/15/probeat-ai-and-quantum-computing-continue-to-collide/

computing might also enable information teleportation amongst other tasks we cannot fathom yet.

All of this remains to be seen, however, the future does not look boring.

# Abbreviations

AI      Artificial Intelligence

ANN    Artificial Neural Network

CS      Computer Science

DL      Deep learning

DS      Data Science

ML      Machine Learning

NN      Neural Network

# References

In order of appearance

1. "cognition" - definition of cognition in English from the Oxford dictionary". www.oxforddictionaries.com. Retrieved 2019-07-02

2. Lowensohn, J. (2015, May 14). Wolfram has created a website that will identify any image you throw at it. Retrieved July 3, 2019, from https://www.theverge.com/2015/5/13/8603531/wolfram-image-identification-site-trained-by-chewbacca

3. (n.d.). Deep Blue. Retrieved July 3, 2019, from https://www.ibm.com/ibm/history/ibm100/us/en/icons/deepblue/

4. Kavraki, L., Svestka, P., Latombe, J., & Overmars, M. (1996). Probabilistic roadmaps for path planning in high-dimensional configuration spaces. IEEE Transactions on Robotics and Automation,12(4), 566-580. doi:10.1109/70.508439

5. Khatib, O. (1986). Real-Time Obstacle Avoidance for Manipulators and Mobile Robots. Autonomous Robot Vehicles, 5(1), 90-98. doi:10.1007/978-1-4613-8997-2_29

6. Safian, R., & Safian, R. (2018, September 11). 5 lessons of the AI imperative, from Netflix to Spotify. Retrieved July 3, 2019, from https://www.fastcompany.com/90234726/5-lessons-of-the-ai-imperative-from-netflix-to-spotify

7. Data Science vs. Computer Science. (n.d.). Retrieved July 4, 2019, from https://www.discoverdatascience.org/articles/data-science-vs-computer-science/

8. Olson, D. L. (2006). Data mining in business services. Service Business,1(3), 181-193. doi:10.1007/s11628-006-0014-7

9. Bishop, C. M. (2016). PATTERN RECOGNITION AND MACHINE LEARNING. Place of publication not identified: SPRINGER-VERLAG NEW YORK.

10. Moroney, Laurence. "Machine Learning Zero to Hero." I/O 2019. Google I/O 19, 9 May 2019, Mountain View, CA, Shoreline Amphitheatre.

11. Oladipupo, T. (2010). Types of Machine Learning Algorithms. New Advances in Machine Learning. doi:10.5772/9385

12. Ayodele, T. O. (2010, February 01). Types of Machine Learning Algorithms. Retrieved July 4, 2019, from http://www.intechopen.com/books/new-advances-in-machine-learning/types-of-machine-learning-algorithms

13. Ullman, S. (n.d.). Unsupervised Learning: Clustering. Lecture presented in Massachusetts Institute of Technology, Cambridge. Retrieved July 4, 2019, from http://www.mit.edu/~9.54/fall14/slides/Class13.pdf

14. Glosser CA. (n.d.). Colored neural network [Digital image]. Retrieved July 07, 2019, from https://commons.wikimedia.org/wiki/File:Colored_neural_network.svg

15. McCorduck, Pamela (2004), Machines Who Think (2nd ed.), Natick, MA: A. K. Peters, Ltd., ISBN 978-1-56881-205-2, OCLC 52197627

16. Nick, Martin (2005), Al Jazari: The Ingenious 13th Century Muslim Mechanic, Al Shindagah, retrieved July 08 2019

17. Berlinski, David (2000), The Advent of the Algorithm, Harcourt Books, ISBN 978-0-15-601391-8,

18. Todhunter, I. (1933). Euclid elements: Books I-VI, XI and XII. Dutton, NY: Everymans Library.

19. Ra⁻shid, R. (2009). Al-Khwa⁻rizmi⁻: The beginnings of algebra. London: Saqi.

20. Buchanan, Bruce G. (Winter 2005), "A (Very) Brief History of Artificial Intelligence" (PDF), AI Magazine, pp. 53–60, archived from the original (PDF) on 26 December 2007, retrieved July 08 2019

21. Ivor Grattan-Guinness (2000) The Search for Mathematical Roots 1870–1940, Princeton University Press, Princeton N.J., ISBN 0-691-05857-1

22. Copeland, P. J. (2012, June 19). Alan Turing: The codebreaker who saved 'millions of lives'. Retrieved July 12, 2019, from https://www.bbc.com/news/technology-18419691

23. Church, A. (1932). "A set of postulates for the foundation of logic". Annals of Mathematics. Series 2. 33 (2): 346–366

24. Crevier, Daniel (1993), AI: The Tumultuous Search for Artificial Intelligence, New York, NY: BasicBooks, ISBN 0-465-02997-3

25. Russell, Stuart J.; Norvig, Peter (2003), Artificial Intelligence: A Modern Approach (2nd ed.), Upper Saddle River, New Jersey: Prentice Hall, ISBN 0-13-790395-2 p.17

26. Ichbiah, D. (2005). Robots: From science fiction to technological revolution. New York: Harry N. Abrams. p. 130

27. Feigenbaum, Edward A.; McCorduck, Pamela (1983), The Fifth Generation: Artificial Intelligence and Japan's Computer Challenge to the World, Michael Joseph, ISBN 978-0-7181-2401-4

28. Moore, Gordon E. (1965). "Cramming more components onto integrated circuits" (PDF). Electronics Magazine. p. 4. Retrieved July 7 2019

29. Markoff, J. (2011, February 16). Computer Wins on 'Jeopardy!' Trivial, It's Not. Retrieved July 11, 2019 https://www.nytimes.com/2011/02/17/science/17jeopardy-watson.html

30. Tascarella, P. (2006, August 14). Robotics firms find fundraising struggle with venture capital shy. Retrieved July 11, 2019, from

https://www.bizjournals.com/pittsburgh/stories/2006/08/14/focus3.html?b=1155528000^1329573

31. Manyika, James; Chui, Michael; Bughin, Jaques; Brown, Brad; Dobbs, Richard; Roxburgh, Charles; Byers, Angela Hung (May 2011). "Big Data: The next frontier for innovation, competition, and productivity". McKinsey Global Institute. Retrieved July 11 2019.

32. Lynch, S. (2017, March 11). Andrew Ng: Why AI Is the New Electricity. Retrieved July 12, 2019, from https://www.gsb.stanford.edu/insights/andrew-ng-why-ai-new-electricity

33. Robitzski, D. (2018, July 18). Artificial intelligence is automating Hollywood. Now, art can thrive. Retrieved July 15, 2019, from https://futurism.com/artificial-intelligence-automating-hollywood-art

34. Lee, K., Dr., & Teddy-Ang, S., Dr. (n.d.). AI Super Powers: China, Silicon Valley and the New World Order. Lecture. Organized by the Lee Kuan Yew School of Public Policy

35. 'Crime-fighter' Jacky Cheung adds 12 crooks, 2 drones to his tally. (2018, September 24). Retrieved July 15, 2019, from https://www.scmp.com/news/china/society/article/2165566/chinas-crime-fighting-pop-star-jacky-cheung-adds-12-crooks-two

36. Tarnoff, B. (2018, October 11). Weaponised AI is coming. Are algorithmic forever wars our future? Ben Tarnoff. Retrieved July 15, 2019, from https://www.theguardian.com/commentisfree/2018/oct/11/war-jedi-algorithmic-warfare-us-military

37. Institute of Public Administration Australia. "In Brief - Artificial Intelligence in the Public Sector". Linked infographic based on information by Daniel Castro, Steve Nichols, Eric Ellis, Cynthia Stoddard (Adobe Chief Information Officer) and Government Technology reporting. Retrieved July 15 2019.

38. CBS News. (2018, April 24). China's behavior monitoring system bars some from travel, purchasing property. Retrieved July 15, 2019, from https://www.cbsnews.com/news/china-social-credit-system-surveillance-cameras/

39. Alexa Terms of Use. (n.d.). Retrieved July 16, 2019, from https://www.amazon.com/gp/help/customer/display.html?nodeId= 201809740

40. Companies that have more money than entire countries. (2018, July 28). Retrieved July 16, 2019, from https://www.rt.com/business/434516-companies-countries-gdp-revenue/

41. Graham-Harrison, E., & Cadwalladr, C. (2018, March 17). Revealed: 50 million Facebook profiles harvested for Cambridge Analytica in major data breach. Retrieved July 16, 2019, from https://www.theguardian.com/news/2018/mar/17/cambridge-analytica-facebook-influence-us-election

42. Jacob Pramuk, J. W. (2019, June 29). US and China agree to continue tariff talks. Here's a timeline of how the trade war started. Retrieved July 16, 2019, from https://www.cnbc.com/2019/06/29/us-china-trade-talks-at-g-20-timeline-of-how-the-tariff-war-started.html

43. Wong, D. (2019, July 03). The US-China Trade War: A Timeline. Retrieved July 16, 2019, from https://www.china-briefing.com/news/the-us-china-trade-war-a-timeline/

44. Boom, D. V. (2018, February 15). Don't use phones from Huawei or ZTE, FBI director says. Retrieved July 16, 2019, from https://www.cnet.com/news/huawei-zte-fbi-chris-wray-nsa/

45. Keane, S. (2019, July 17). Huawei ban: Full timeline on how and why its phones are under fire. Retrieved July 17, 2019, from https://www.cnet.com/news/huawei-ban-full-timeline-on-how-and-why-its-phones-are-under-fire/

46. Kharpal, A. (2019, May 08). Huawei CFO's extradition case: Everything you need to know. Retrieved July 19, 2019, from

https://www.cnbc.com/2019/05/08/huawei-cfo-meng-wanzhou-extradition-case-everything-you-need-to-know.html

47. What is 5G?: Everything You Need to Know About 5G. (2019, April 12). Retrieved July 19, 2019, from https://www.qualcomm.com/invention/5g/what-is-5g

48. Fisher, T. (2019, July 03). 5G vs 4G: Everything You Need to Know. Retrieved July 19, 2019, from https://www.lifewire.com/5g-vs-4g-4156322

49. Wu, Debby, et al. "Who's Winning the Tech Cold War? A China vs. U.S. Scoreboard." Bloomberg.com, Bloomberg, 20 July 2019, www.bloomberg.com/graphics/2019-us-china-who-is-winning-the-tech-war/.

50. Vincent, James. "A Never-Ending Stream of AI Art Goes up for Auction." The Verge, The Verge, 5 Mar. 2019, www.theverge.com/2019/3/5/18251267/ai-art-gans-mario-klingemann-auction-sothebys-technology.

51. Bergland, C. (2013, September 12). The "Love Hormone" Drives Human Urge for Social Connection. Retrieved July 22, 2019, from https://www.psychologytoday.com/us/blog/the-athletes-way/201309/the-love-hormone-drives-human-urge-social-connection

52. Wolla, S. A., Dr., & Sullivan, J. (2017). Education, Income, and Wealth (Rep.). St. Louis, Missouri: Federal Reserve Bank of St. Louis. doi:research.stlouisfed.org Accessible at: https://files.stlouisfed.org/files/htdocs/publications/page1-econ/2017-01-03/education-income-and-wealth_SE.pdf

53. Dalio, R., Kryger, S., Rogers, J., & Davis, G. (2017, March 22). Populism: The Phenomenon. Retrieved July 26, 2019, from https://www.bridgewater.com/resources/bwam032217.pdf

54. Piech, C. (2013). K Means. Retrieved July 24, 2019, from https://stanford.edu/~cpiech/cs221/handouts/kmeans.html

55. K-means clustering algorithm - Data Clustering Algorithms. (n.d.). Retrieved July 25, 2019, from https://sites.google.com/site/dataclusteringalgorithms/k-means-clustering-algorithm

56. Rahul. (2018, August 08). Learn ML Algorithms by coding: Decision Trees. Retrieved July 26, 2019, from https://lethalbrains.com/learn-ml-algorithms-by-coding-decision-trees-439ac503c9a4

57. Illustration from https://iq.opengenus.org

58. Weisstein, Eric W. "Acyclic Digraph." From MathWorld--A Wolfram WebResource. http://mathworld.wolfram.com/AcyclicDigraph.html

59. Illustration by David Eppstein (2013)

60. Wylie, C. (2019, May 15). 5 Essential Neural Network Algorithms. Retrieved July 27, 2019, from https://opendatascience.com/essential-neural-network-algorithms/?source=post_page--------------------------

61. Goodfellow, I., Bengio, Y., & Courville, A. (2017). Deep learning. Cambridge, MA: MIT Press.

62. Jadon, S. (2018, March 19). Introduction to Different Activation Functions for Deep Learning. Retrieved July 28, 2019, from https://medium.com/@shrutijadon10104776/survey-on-activation-functions-for-deep-learning-9689331ba092

63. Information and Likelihood Theory: A Basis for Model Selection and Inference. (n.d.). Model Selection and Multimodel Inference, 49-97. doi:10.1007/978-0-387-22456-5_2

64. Bande, A. (2018, March 18). What is underfitting and overfitting in machine learning and how to deal with it. Retrieved July 28, 2019, from https://medium.com/greyatom/what-is-underfitting-and-overfitting-in-machine-learning-and-how-to-deal-with-it-6803a989c76

65. Turing, A. M. (2016, December 23). London Mathematical Society Journals. Retrieved July 28, 2019, from

https://londmathsoc.onlinelibrary.wiley.com/doi/abs/10.1112/plms/s2-42.1.230

66. AlphaGo. (n.d.). Retrieved July 28, 2019, from https://deepmind.com/research/alphago/

67. Markov, A. A., & Schorr-Kon, J. J. (1968). Theory of algorithms. Jerusalem: Israel Program for Scientific Translations.

68. Acccessible here: https://towardsdatascience.com/self-learning-ai-agents-part-i-markov-decision-processes-baf6b8fc4c5f

69. Moore, G. (1998). Cramming More Components Onto Integrated Circuits. Proceedings of the IEEE, 86(1), 82-85. doi:10.1109/jproc.1998.658762

70. Protalinski, E. (2019, March 15). ProBeat: AI and quantum computing continue to collide. Retrieved July 29, 2019, from https://venturebeat.com/2019/03/15/probeat-ai-and-quantum-computing-continue-to-collide/

Disclaimer:

Although the author has made every effort to ensure that the information in this book was correct at press time, the author does not assume and hereby disclaims any liability to any party for any loss, damage, or disruption caused by errors or omissions, whether such errors or omissions result from negligence, accident, or any other cause.

To contact the author, please go to
www.edge-ai.net

www.ingramcontent.com/pod-product-compliance
Lightning Source LLC
Chambersburg PA
CBHW041106180526
45172CB00001B/127